LET'S LOOK AT LANDFORMS

THOSE ARE HILLS!

I0794175

BY DWAYNE HICKS

Gareth Stevens PUBLISHING

Please visit our website, www.garethstevens.com. For a free color catalog of all our high-quality books, call toll free 1-800-542-2595 or fax 1-877-542-2596.

Library of Congress Cataloging-in-Publication Data
Names: Hicks, Dwayne, author.
Title: Those are hills! / Dwayne Hicks.
Description: New York : Gareth Stevens, [2022] | Series: Let's look at landforms | Includes index.
Identifiers: LCCN 2020032261 (print) | LCCN 2020032262 (ebook) | ISBN 9781538263839 (library binding) | ISBN 9781538263815 (paperback) | ISBN 9781538263822 (set) | ISBN 9781538263846 (ebook)
Subjects: LCSH: Mountains–Juvenile literature.
Classification: LCC GB512 .H54 2022 (print) | LCC GB512 (ebook) | DDC 551.43/6–dc23
LC record available at https://lccn.loc.gov/2020032261
LC ebook record available at https://lccn.loc.gov/2020032262

Published in 2022 by
Gareth Stevens Publishing
29 E. 21st Street
New York, NY 10010

Copyright © 2022 Gareth Stevens Publishing

Designer: Tanya Dellaccio
Editor: Greg Roza

Photo credits: Cover Stewart Hall/iStock/Getty Images Plus/Getty Images; pp. 3–22 (background texture) MicroOne/Shutterstock.com; p. 5 Dennis Fischer Photography/Moment/Getty Images; p. 7 George W Johnson/Moment/Getty Images; p. 9 instamatics/E+/Getty Images; p. 11 Arie Quick/EyeEm/Getty Images; p. 13 Steffen Schnur/Moment/Getty Images; p. 15 CHENG FENG CHIANG/iStock/Getty Images Plus/Getty Images; p. 17 Dave Moorhouse/Moment Open/Getty Images; p. 19 MattGush/iStock/Getty Images Plus/Getty Images; p. 21 BJI/Blue Jean Images/Getty Images.

All rights reserved. No part of this book may be reproduced in any form without permission in writing from the publisher, except by a reviewer.

Printed in the United States of America

CPSIA compliance information: Batch #CSGS22: For further information contact Gareth Stevens, New York, New York at 1-800-542-2595.

Find us on

CONTENTS

Boldface words appear in the glossary.

Hello, Hills!

A hill is an area of land that's higher than the areas around it. Hills have **sloping** sides. Some hills are tall and **steep**. Others are flatter. Some hills are covered in trees and grass. Others are rocky.

Mountain or Hill?

Mountains are taller than hills. They are often steeper too. People around the world have different ideas about how big a landform needs to be to call it a mountain. A hill in one country might be called a mountain in another country.

Between the Hills

Hills come in many sizes. Large hills are often surrounded by lower areas called valleys. Valleys often have rivers running through them. This makes valleys a great place for people to build towns and cities.

Building Up

Earth's **surface** is made of pieces called "plates." They move very slowly and can bump into each other. This can cause some areas to rise higher than others. These areas become hills. If they keeping rising, they can become mountains!

Wearing Down

Erosion creates hills. Wind and water wear away at the land. After many years, this can cause a mountain to grow smaller. In time, the mountain may become a hill. Rivers can wear away at flat land and make hills.

Building Up Again!

Erosion wears away at land. Wind and water can carry small bits of rock and sand away from one place and leave it in another place. As the rock and sand pile up, they create new landforms, such as hills.

Left Behind

Glaciers are huge masses of ice that form in cold mountains. They get bigger over time. A glacier "flows" very slowly. It picks up rock as it moves. When a glacier **melts**, it leaves behind the rock. This makes hills.

Man-Made Hills

Hills made by people are called mounds. Some Native American groups built mounds. They were very special places to them. Modern **mining** causes a lot of waste rock. This rock is dumped in piles that form tall hills.

MONKS MOUND
COLLINSVILLE, ILLINOIS

People and Hills

People have long lived on and around hills. People built towns on hills to avoid **floods**. Many towns and cities were built on hills. Some hills are great places for hiking and hunting. Some are perfect for sledding and skiing!

GLOSSARY

erosion: the wearing away of the earth by wind or water

flood: a large amount of water covering an area of land that is usually dry

melt: to change from a solid (such as ice) to a liquid (such as water) when heated

mining: digging into the earth to reach metals and other objects

sloping: having a slanted side like a hill

steep: almost straight up and down

surface: the outer layer of something

FOR MORE INFORMATION

BOOKS

Amstutz, Lisa J. *Hills.* North Mankato, MN: Capstone, 2020.

Hutchison, Patricia. *Valleys.* Mendota Heights, MN: North Star Editions, 2018.

WEBSITES

Landforms

jr.brainpop.com/science/land/landforms/

BrainPop Jr. has an interactive website with information about all kinds of landforms.

Landforms | PBS LearningMedia

ny.pbslearningmedia.org/resource/buac17-k2-sci-ess-landforms/landforms/

Watch videos about different landforms, including hills.

Publisher's note to educators and parents: Our editors have carefully reviewed these websites to ensure that they are suitable for students. Many websites change frequently, however, and we cannot guarantee that a site's future contents will continue to meet our high standards of quality and educational value. Be advised that students should be closely supervised whenever they access the internet.

INDEX